2/01

Animal Therapist

WEIRD
CAREERS
in SCIENCE

Animal Therapist

Kay Frydenborg

CHELSEA HOUSE
PUBLISHERS
A Haights Cross Communications Company ®
Philadelphia

CHELSEA HOUSE PUBLISHERS

VP, New Product Development Sally Cheney
Director of Production Kim Shinners
Creative Manager Takeshi Takahashi
Manufacturing Manager Diann Grasse
Series Designer Takeshi Takahashi
Cover Designer Takeshi Takahashi

Staff for **ANIMAL THERAPIST**

Project Management Ladybug Editorial and Design
Development Editor Tara Koellhoffer
Layout Gary Koellhoffer

A Haights Cross Communications ◢◤ Company ®

www.chelseahouse.com

First Printing

9 8 7 6 5 4 3 2 1

Library of Congress Cataloging-in-Publication Data

Frydenborg, Kay.
  Animal therapist / Kay Frydenborg.
    p. cm. — (Weird careers in science)
  Includes bibliographical references and index.
  ISBN 0-7910-8704-2
  1. Animals—Therapeutic use—Juvenile literature. I. Title. II. Series.
  RM931.A65F79 2005
  615.8'515—dc22
                              2005012071

TABLE OF CONTENTS

Chapter 1

A Day in the Lives of Three Animal Therapists

DOG ON DUTY

A TALL, ELEGANT GREYHOUND walks calmly along the gleaming white corridor of a hospital ward beside her owner as nurses, doctors, and family members hurry past. Her sleek brindle coat gleams in the fluorescent light and her neatly clipped toenails click on the tile floor like high-heeled shoes. Progress is slow, because every few steps someone stops to admire the big dog or ask her name. Each time this happens, her tail wags slowly and she stands patiently while a busy nurse or housekeeper pats her gently on top of her long, narrow head. Finally, the dog's owner knocks softly on the door of a patient's room. Nurses have told the dog's owner that the patient is an elderly man who has suffered a stroke that has left him partially paralyzed. Without hesitation, the dog and her owner approach the man's bed, where he lies propped up on pillows, staring blankly into space.

"Mr. Johnson?" says the woman with the dog. "This is Molly. Would you like to visit with her?" The man turns toward the woman and the dog, smiles faintly, and nods his head. Molly's owner speaks softly to the dog, and the animal walks straight to Mr. Johnson's bedside. At first, the man seems confused, but then Molly gently lays her big head on his shoulder and half-closes her eyes. Mr. Johnson reaches out and begins softly stroking the dog's head. In a moment, he is running his wrinkled hand all the way down Molly's long back, even lifting himself slightly out of the bed to reach the tip of her tail. He tries to speak, but the words come out garbled. Still, the smile on his face and the tears welling in his eyes show clearly that petting Molly is a very emotional experience.

After a few minutes, the man is starting to get tired. Molly gives him a soft dog kiss on his chin, and her owner says good-bye, promising to return next week if the man is still in the hospital. Mr. Johnson is already sleeping peacefully by the time the woman with the dog taps softly on the door of the next patient's room.

HEALING HORSES

The young girl walks into the weathered barn under her own power, but it's a slow process as her crutches sink into the soft, sandy soil of the path. The look of determination on her face shows that the girl is used to the struggle to move. Her mother follows a few steps behind, slowing her own pace to match her daughter's, clasping her hands behind her back as if to force herself not to help if she is not truly needed. Once inside the dimly lit barn, a sort of lightness and anticipation come over both the girl and her mother. Babe, a big gray mare, is waiting.

The girl knows what to do with the brushes and combs packed neatly in a small box inside the tackroom. Babe stands still, her solid bulk half-supporting the girl's small body. The little girl gently strokes the horse's soft, white coat and inhales the special horse aroma that she has come to love. Before long, with help from her teacher and some volunteers, the girl is sitting way up high on a specially made saddle that has been fastened onto Babe's broad back. She is taller than anyone now. She giggles with joy when Babe reaches around to nuzzle her foot. Then, after a quiet prompt from her teacher, she says in a small, clear voice, "Walk on, Babe!" Babe obeys.

The girl was born with **cerebral palsy**. She has trouble walking, and her speech is sometimes hard to understand. But she loves horses more than just about anything now, and each week that she rides Babe, she grows stronger and her muscles get more relaxed. Her doctor is amazed at her progress, and says that he believes she may soon be able to walk without any crutches at all. She may even be able to go to a regular school one day, because talking to Babe has improved her speech so much.

The girl doesn't think about these things, though, during the long week between the days when she gets to ride Babe. What she thinks and dreams about is rocking along on the back of the big, gray horse. She feels strong when she rides, unlike most other times in her life, and she feels proud that she can control an animal as big as Babe (Figure 1.1). It's

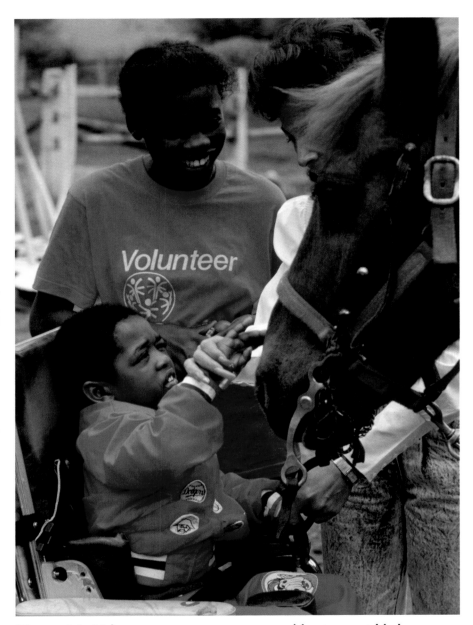

Figure 1.1 Volunteers encourage a young rider to greet his horse during a therapeutic riding session.

something that not many people—not even her mother or her father or her big brother, who sometimes teases her for her slowness—can do. Her teacher gives her exercises to do on horseback. They play games, like riding backward and reaching way out to touch Babe's tail. Sometimes, her teacher asks her to reach down and touch her own toes, or to stretch her stiff arms all the way up on Babe's neck, to place a plastic ring over the horse's ear. Babe sometimes flicks her ear as if the ring tickles, but she never seems to mind. When the riding is over, before the girl takes up her crutches again for the tough walk back to her mother's car, she plants many kisses on the whiskery nose of the gray mare.

DOLPHIN DOCTORS

Bright sunlight glints off the glassy surface of a Florida lagoon, where a boy named Aubrey drifts through blue-green water, suspended by a bright yellow flotation vest. Nearby, Aubrey's therapist, a young man named Chris, floats quietly while keeping a close watch on Aubrey. They do not speak, or even touch. Aubrey hates to be touched, and he hardly ever speaks. He has **autism**—a type of brain disorder that affects his speech and movement, and makes it difficult for him to focus on and relate to other people. Sometimes he will say words when Chris shows him cards with pictures of various objects and animals, but even then he will gaze around the room, not looking at his teacher or thinking about the lesson at hand. He avoids looking people in the eye, even his parents, and it is hard for anyone to hold his attention. But he loves the water, which seems to soothe him when he feels agitated or upset. When Chris promises him a chance to get into the water, Aubrey is often willing to concentrate harder on his speech therapy and other exercises than he ever did before he began to swim.

Today has been a good day in the classroom, and now it is time for Aubrey's reward. As the boy floats along, nodding his head and humming softly to himself, Chris sees a graceful, massive gray body gliding up, just under the surface of the water.

"Here's Sarah, Aubrey!" he says. In the next moment, the dolphin's great, bottle-shaped snout pops out of the lagoon, and she gives Aubrey a toothy grin as she holds all eight feet of her sleek body nearly vertical in the salty water, just inches from Aubrey's astonished face.

"Look, Aubrey, Sarah's come to say hello to you!" calls Chris encouragingly. Then, as the dolphin rolls sideways in the water without making a splash, he adds, "Now she's asking you to rub her belly! Do you want to?" After a moment's hesitation, the boy reaches his right hand toward Sarah, and, with one finger, strokes her hairless, rubbery side.

"Very *good*, Aubrey!" says Chris. "What does it feel like? Maybe like a hard-boiled egg, after you take the shell off?" Aubrey doesn't answer, but his attention is riveted on the big bottlenose dolphin now swimming slowly around him in a close circle. When she stops in front of him and peers into his eyes with her own, the boy returns her gaze. He has never looked into the eyes of any human for more than a second or two, but now he looks into the eyes of Sarah the dolphin for what seems like minutes before looking away. By this time, Chris, his therapist, has paddled over close to the boy.

"Aubrey," he says, "I think today is a perfect day for you to swim with a dolphin, don't you?"

"Swim," says Aubrey. Chris shows him how to gently take the dolphin's **dorsal** fin in his hand and hold on, as

Sarah starts gliding in an easy circle around the lagoon. She goes slowly at first, but steadily gains speed while supporting the boy's body along the surface of the water. When she comes to a gentle stop, Aubrey is laughing with joy.

Prescription Pets?

Americans love animals. In fact, the rate of pet ownership in the United States is four times higher than that in Europe, and five times higher than that in Japan. In study after study, Americans reveal their deep attachment to their animals, with up to 90% of pet owners claiming that their pet is "extremely important" to them, and close to 80% identifying a pet as their closest companion. Animal images are everywhere: in art, literature, music, food, language, and religion. They fill our imagination and our dreams. According to a 1983 study, up to 57% of the dreams of 4-year-old boys involve animals.

Now more than ever, animals are a major focus of scientists and health care workers. Domestic animals and livestock are now recognized as a potential "early warning system" against the post–9/11 threat of **bioterrorism**. In the emerging field of animal-assisted therapy, some animals have a new role to play—"prescription medicine" to cure a long list of human ills.

Animal therapists, like all professionals, have specific tools that help them do their work. But unlike most science-related professions, in animal-assisted therapy, the main tool is a living, breathing creature. It doesn't live in a laboratory or a test tube, but more often in a kennel, a stable, or someone's home, and it probably has a name and personality all its own.

ANIMAL THERAPISTS

What do all three of these animals have in common? Besides the fact that all of them are **mammals**, the most important thing is that Molly, Babe, and Sarah are all part of a growing field called **animal-assisted therapy (AAT)**. The humans who work with these animals to help people with special needs are **animal therapists**. The term can be confusing, because an animal therapist is not a therapist *for* animals, like a veterinarian—although the health and welfare of the animals they work with are very important to them. An animal therapist is a health-care professional who uses animals to help treat *people* with various mental and physical disabilities, or to help people who are dealing with abuse, stress, neglect, or other problems.

Animal therapy is a diverse field. It can involve many different kinds of animals, and trained professionals from a variety of health-related disciplines. It can be used in all types of settings, from hospitals and nursing homes to schools, prisons, and residential treatment facilities. There are also plenty of opportunities for volunteers who love animals.

What all animal therapists share is a love and concern for both animals and people, a curious mind, and a desire to help people who may be confused, depressed, or in pain. Sometimes, for these people, an animal can turn out to be the best medicine of all.

Chapter 2

What Is Animal Therapy?

ANIMALS IN HISTORY

HUMAN BEINGS HAVE BEEN OBSERVING, hunting, using, and bonding with animals since prehistoric times. The world's oldest known paintings, drawn by unknown artists on the walls of the Chauvet cave in France more than 30,000 years ago, depict horses, rhinoceroses, lions, buffalo, and mammoths in remarkable detail. Of all the subjects that could have been chosen, it was animals that were important enough to be represented by the painstaking work of ancient artists.

Animals have always made people feel better or stronger in many ways, but no one knows precisely when they were first used for **therapeutic** purposes, to heal human illness. Some people may have thought of animals this way from the very beginning of time. An account from the 9th century in Gheel, Belgium, mentions animals being included in what was called *"therapie naturelle."* This was a progres-

Therapy Dogs of the Ancient World

Healing powers have been attributed to dogs for as long as dogs have been associated with humans. In ancient Egypt, the dog-headed god Anubis was physician to the gods and guardian of the mysteries of making mummies and **reincarnation**. The Sumerian goddess Gula the Great Physician used the image of a dog as her sacred emblem, as did Marduk, the Babylonian god of healing and reincarnation.

In Greek mythology, Askelepios, son of Apollo and the God of medicine, established his shrine in a sacred grove—a sort of ancient health spa. The afflicted went there to seek cures for a range of ailments. Treatment involved various rites of purification and sacrifice, followed by periods of possibly drug-induced sleep within the shrine. While they slept, patients were visited by the god, often in the guise of a dog that licked the injured part of the sleeping person's body. Meanwhile, real dogs that lived at the shrine were specially trained to lick the people with great affection. It was widely believed that the dogs were representatives of the god himself, and that they had the power to cure illness with their tongues. Inscribed tablets found at the temple record the miraculous powers of these dogs in entries like this one:

Thuson of Hermione, a blind boy, had his eyes licked in the daytime by one of the dogs about the temple, and departed cured.

sive community program through which local citizens cared for handicapped people. Household pets and farm animals played a central part in the program. In the 1700s, horses were being used to treat various diseases, though detailed accounts of early "**hippotherapy**" are scarce. Even back then, however, people with **neurological** disorders achieved better balance and enhanced motor abilities through horseback riding.

The first specific reports of animal-assisted therapy (AAT) came from the York Retreat, founded in England in 1792 by the Society of Friends, or Quakers. This institution for people with severe mental health problems was based on the idea that animals would enhance the "humanity" of the emotionally ill. Patients were treated with kindness and respect—a revolutionary idea at the time—and were encouraged to care for rabbits, chickens, and other farm animals. It was believed that people who seemed "out of control" could develop self-control by caring for creatures that were weaker than themselves.

During the Victorian Era of the mid- to late 1800s, public criticism of the appalling conditions in **asylums** and prisons led to a wider use of pets as part of an effort to humanize these institutions. In 1860, famous British nurse Florence Nightingale observed that a small pet "is often an excellent companion for the sick, for long **chronic** cases especially." In 1867, pets played a role in the treatment of **epileptics** at Bethel, in Bielefeld, Germany.

In the United States, the first recorded use of animals in therapy occurred in 1942 at an Army Air Corps Convalescent Hospital in Pawling, New York. Animals were found to provide health benefits that other forms of medical treatment could not.

MODERN ANIMAL THERAPY

The modern profession of animal-assisted therapy was born in 1962, when a New York **psychotherapist** named Boris Levinson described how his dog, Jingles, had helped him better communicate with children in his practice. Levinson had discovered the positive effect Jingles had on young patients by chance: One day, a very withdrawn boy arrived early for his appointment, and the dog was in Levinson's office. Levinson immediately noticed that the child responded well to Jingles, and from that day on, the young patient was less distant and more inclined to talk during his therapy session, as long as Jingles was in the room. It seemed to Levinson that the dog was able to act as a sort of go-between, helping the child feel more at ease in the strange and somewhat frightening therapy setting. Although Levinson's claims were initially ridiculed by his fellow professionals (who often asked jokingly if Levinson shared his fee with the dog), the concept of using animals to help in the **psychological** treatment of children began to win support. When Levinson surveyed 435 psychotherapists in New York State in 1972, he found that one-third of them had used pets in their practices.

During the early 1970s in Ann Arbor, Michigan, the Children's Psychiatric Hospital adopted a resident dog name Skeezer, who spent seven years on the ward, "proving that with proper training a dog can help open pathways into the minds and hearts of disturbed children." Skeezer was so popular that after he retired, he became the subject of a book and a TV movie. By this time, "petmobile" programs had begun to spring up around the country. They would bring visiting animals to nursing homes and other institutions for people with special needs.

The field of animal-assisted therapy has continued to grow through the present day. It has now expanded to include "assistance animals" like guide dogs for the blind and hearing dogs for the deaf. In 1989, the Delta Society was created to oversee this rapidly expanding field in the United States, and a Delta Society–sponsored program called "Pet Partners" was started to provide training standards and guidelines for certifying therapy and assistance animals and their handlers nationwide. At the same time, therapeutic horseback riding programs were developed by the North American Riding for the Handicapped Association (NARHA).

BUT IS IT SCIENCE?

Not all, or even most, animal therapists are scientists. After four decades of research and experience in the field, though, there is plenty of evidence that the basic **hypothesis** of animal therapy—that the human-animal bond can provide both **physiological** and mental benefits for human health— is firmly rooted in **anthropology**, **psychology**, **evolutionary biology**, medicine, and many other scientific disciplines, as well as fields like history, social science, literature, mythology, and even religion.

Some animal therapists are both **practitioners** and researchers. Many work at medical and veterinary schools, and others come to animal therapy after previous careers as teachers, animal trainers, and health-care professionals. Animal therapy is truly an **interdisciplinary** approach to human health.

Many studies have focused on the ways pets and animals in general contribute to human mental health, but the conclusions of researchers have often been largely **subjective**, rather than founded in hard scientific data. That is, their findings are based on people's personal accounts of how animals have improved their lives, or on the observations of healthcare workers whose own beliefs may influence their conclusions. Even so, much of this evidence is very persuasive: Animals have been found to help fend off loneliness and depression, to give people a sense of safety and protection, and to encourage physical activity and social interaction in people who would otherwise spend most of their time alone. For some people, an animal may serve as a substitute for absent children, or as a welcome distraction from pain and trouble. Pets are a source of amusement and companionship, and a socially acceptable outlet for touching and caressing, which many studies have shown is as basic a human need as food. Pets provide a sense of order and structure to people's lives, and allow them to feel needed, appreciated, and unconditionally loved. They are ready partners in play, another basic human activity shared with many other animals.

Some experts also believe that we humans crave a connection with the natural environment, something that can be hard to come by in a time of increasing urban sprawl and fast-paced, high-tech lifestyles. In a sense, animals may be our modern ambassadors to the natural world. They answer a widespread longing that many of us may feel to restore a kind of simplicity to our hectic lives.

However widespread and longstanding the belief that animals are good for us, hard science to support the benefits of animal therapy has been a long time coming. Part of the reason is that it is difficult to measure psychological benefits. We usually think of mental health and physical health as two separate things, but most scientists and health researchers now believe there is no clear dividing line between the mind and the body. Mental health problems and stress often lead to physical illness; physical illness, in turn, can have serious consequences for a person's mental health.

Besides the well-known psychological benefits of human interaction with animals, scientific studies have now shown numerous physiological benefits, including:

- higher survival rates following diagnosis of heart disease or heart attacks;
- reduced blood pressure and stress levels, plus lower cholesterol levels;
- better balance, coordination, mobility, muscular strength, posture, and language ability for physically challenged people who take part in therapeutic riding programs;
- a dramatic increase in recreational walks, which improves physical fitness;
- a reduction in the demand for physicians' services for routine medical problems.

Animals have been credited with some pretty dramatic "cures," such as bringing people out of **comas** and prompting autistic children to speak for the first time ever. No longer are such stories confined to the realm of fiction, although animal healers continue to play starring roles in popular novels, movies, and television shows, as more and more real-life instances of animal healing are documented and analyzed by scientists from many different fields.

SOME IMPORTANT TERMS TO KNOW

What are the proper terms to use in describing the animals used in human health-care settings? Standard definitions promoted by the Delta Society, which now oversees the field of animal therapy in the United States, replace the word *pet* with *animal*, since not every animal used in therapy is, strictly speaking, a "pet." The current definitions include:

- **Animal-assisted Activities (AAA)** provide opportunities for motivational, educational, and/or recreational benefits to enhance quality of life, but are not necessarily provided by a licensed professional therapist. AAA specialists may or may not have specialized training, and may come from a variety of backgrounds such as educators, riding instructors, dog trainers, nurses, physical therapists, 4-H leaders, or trained volunteers.

- **Animal-assisted Therapy (AAT)** is a more focused, goal-directed treatment program in which an animal is an integral part of the treatment process. AAT is delivered or directed by a trained and licensed health/human service provider in a variety of settings to both groups and individuals, and is rigorously documented and evaluated. AAT specialists may be speech pathologists, psychologists, social workers, physicians, licensed counselors, or other health-related professionals.

- **Human-animal Support Services** may be professionals or trained volunteers working to support a team of people and animals, and to facilitate those relationships. They may be animal behaviorists, animal trainers, grief counselors, army/police/rescue K-9 unit handlers, or pet foster-care providers. Although many have professional training, they may be volunteers.

Figure 2.1 Dogs are often used in prisons to help inmates learn responsibility and how to care for others.

WHERE ANIMAL THERAPISTS WORK

We're all familiar with animals in the home. Altogether, Americans own more than 500 million pet animals, including 52.9 million dogs and 59.1 million cats. Pet owners spend billions of dollars a year to feed and care for their animals—more money is spent on pet food than on baby food—and it's safe to say that the relationship of people to their own pets is often therapeutic, even if we don't usually think of it that way.

But today, animals are also found in many different institutional settings, including nursing homes and **hospices**, hospitals, schools, prisons, and residential and outpatient programs for "at risk" youth (Figure 2.1). Almost anywhere people who are troubled or ill receive treatment and support, animals are a part of the program.

Chapter 3

Angels on a Leash

JOSH, A BIG, BLACK, HUGGABLE bear of a dog, is a superstar. Winner of the coveted "Best in Show" title at the Westminster Kennel Club's 128th annual dog show in February 2004, the 155-pound Newfoundland (Figure 3.1) bounded into the limelight as "America's Dog." He made appearances on popular TV shows like *Good Morning America* and *Late Night with David Letterman*, and was featured in editorial cartoons in several national newspapers. His hometown of Flemington, New Jersey, even proclaimed June 14, 2004, "Josh Day." Hundreds of fans and several state government officials turned out that June

Figure 3.1 Josh struts his stuff in competition before beginning a new career as a therapy dog.

day to honor him, admirers lined up for photos, and the city's mayor presented him with the "bone to the city."

But Josh didn't let all the fanfare go to his head. By November 2004, now retired from the show ring, the 4-year-old was beginning his new career as a certified therapy dog. On November 9, the media-savvy canine headlined an event at Morgan Stanley Children's Hospital of New York–Presbyterian, where he served, along with five other therapy pooches, as "spokesdogs" for the hospital's new animal therapy program. In partnership with the Westminster

Kennel Club, the program is dubbed "Angel on a Leash." At the ceremony, hospital executive director Cynthia Sparer related the mission of Angel on a Leash: "Pets are often a dearly missed part of the family when children require extensive stays at the hospital," she said. ". . . Angel on a Leash will provide an opportunity for our children to play, be comforted, and express themselves, just as they would at home." After the public event, during which the dogs met and greeted crowds of fans and paparazzi, Josh led a group to the sixth floor of the hospital to visit young patients who were too ill to attend the ceremony.

HELPING AFTER TRAGEDY

On May 21, 1998, a troubled 15-year-old student opened fire on his classmates in the cafeteria at Thurston High School in Springfield, Oregon, killing two people and wounding 25. Many students and teachers were deeply traumatized by the event; the school provided counseling to all who needed it. Among those who answered the call for help were Garth, a golden retriever, and Bear, a keeshond. They and their owners are members of a nationwide volunteer organization called Pet Partners, which coordinates animal visits to hospitals and nursing homes. This was the first time they'd been called to help in a crisis. Every day for a week, the two teams—Sandi Arrington and Garth, and Cindy Ehlers and Bear—spent time with students and faculty at the school. The dogs were there for petting or play, and responded with affection to the students who approached them. Sometimes the dogs even went up to students on their own and "asked" to be petted.

"Some students just put their arms around the dogs and wept into their fur," said Arrington. "The dogs seemed to understand that something serious and tragic had happened. Also, they contributed an air of normalcy to the situation."

In the aftermath of the terrorist attacks of September 11, 2001, Ehlers responded once again, this time bringing her 2-year-old keeshond, Tivka, to New York City. There, the two joined grieving families of victims at Ground Zero. As it turned out, therapy dogs were as much a comfort to overwhelmed rescue workers as they were to families.

"Over and over again," said Ehlers. "I've heard men say, 'That dog made my day.'"

HELPING TROUBLED CHILDREN

As you read in Chapter 2, in 1962, New York psychotherapist Boris Levinson made a surprising discovery when his dog Jingles bonded with a very withdrawn young patient. Later, Levinson explained how this dog-centered therapy seemed to work:

> . . . All children have an intense need to master someone or something that does not talk back, that accepts one regardless of what one is. This is overwhelmingly prevalent among disturbed children who especially do not want to be judged. . . . Disturbed children have a strong need for physical contact but are afraid of human contacts because they have been hurt so much and so often by people. Since the hurt is not associated with the dog, this conflict resolves itself. They will permit a dog to approach them and they will pet the animal while telling him all about their difficulties. A dog apparently poses less of a threat. . . .

By chance, Levinson and Jingles had laid the foundation for animal-assisted therapy as we know it today.

THERAPY DOGS AND SERVICE DOGS

The examples already cited are just a few of the ways therapy dogs work today. They assist in both physical and mental health therapies for all kinds of patients. Many people may confuse *therapy* dogs with *service* dogs. Though related, the terms are not synonymous. **Service dogs** are trained to work individually with a disabled person, performing tasks the person cannot do easily. Along with a few other service animals, service dogs are defined and covered under the 1990 Americans With Disabilities Act, which protects the rights of people with disabilities to have their service animals with them in public places, transportation vehicles, and workplaces. Service dogs have been trained to do all kinds of jobs, to help people with impairments in walking, breathing, seeing, hearing, or otherwise caring for themselves.

Perhaps the best-known type of service dog is the guide dog for the blind (Figure 3.2). But dogs have also been trained to fetch dropped or needed items; to pull wheelchairs; to turn lights on and off; to open elevator doors; to hand paperwork to receptionists; to carry things in special backpacks; to pick up items off a shelf; or to alert a person to visitors, ringing phones, and emergencies, among many other tasks. Some dogs are trained to provide support for a person who needs help walking, and others signal epileptic individuals before a seizure occurs, so the person can avoid dangerous falls. Scientists don't yet know for sure how some dogs are able to sense that a person will soon have a seizure. Dogs may be able to detect subtle changes in human body language, expression, or behavior, or respond to changes in body odor triggered by abnormal nervous system activity in an epileptic person.

In addition to performing these essential tasks, of course,

service dogs provide the people they assist with social and emotional support. Many disabled individuals consider their service dog their best friend. Some insist they would never give up their dog, even if they could get rid of their disability in the bargain.

PRISON PUPS

Another way the line can be blurred between service dogs and therapy dogs is illustrated in programs like Project POOCH or Puppies Behind Bars. In these and other similar programs, inmates and juvenile offenders are given the opportunity to train homeless dogs for future careers as service dogs, therapy dogs, or police dogs. This provides both the animal and inmate with important skills for the future, while being therapeutic for both.

Project POOCH (Positive Opportunities—Obvious Change with Hounds) is an Oregon program that trains juvenile offenders to care for homeless dogs from a local humane society. The dogs, which risk being **euthanized** if they don't find homes, are trained by kids in the program to make the animals more adoptable. The student trainers work with their dogs daily. In the process, the trainers themselves learn to be responsible, reliable, and patient. They experience, perhaps for the first time in their lives, a sense of being needed and appreciated, of making a difference in the life of another living thing. By managing their dogs, the kids learn how to manage their own behavior. Those who have been making excuses for their behavior discover that dogs don't care about excuses. As part of the program, kids also earn school credits, develop good work habits, and gain valuable job skills.

Experiencing unconditional love for the first time helps

Figure 3.2 A guide dog makes crossing a busy street much safer for someone who is blind or visually impaired.

both students and dogs develop self-confidence and the positive attitude they'll need to build future relationships. As Chris, a participant in Project POOCH, learned while working with Ginger, an abused and abandoned English pointer, "Trust is easy to lose and hard to get back." There's ample proof that this kind of program works: Not a single youth from Project POOCH has been involved in any further criminal activity, and the staff members working with the students attest to great improvements in their respect for authority, social interaction, and leadership skills. Similar programs around the country have noted equally positive outcomes.

Puppies Behind Bars trains adult women inmates to raise puppies for eventual training as guide dogs for the blind. The program began at New York State's maximum-security Bedford Hills Correctional Facility and eventually expanded to include five prisons, with about 50 puppies in training at any given time. The puppies live in the cells with their inmate caretakers, attend classes once a week, and spend two or three weekends a month with puppy sitters outside the prison, so they can be exposed to the outside world. After 16 months, the pups are tested to see if they'll make the grade as guide dogs. If so, they return to their original guide dog schools for further training. Those that aren't qualified are donated to families with blind children. Either way, the dogs will spend their lives as companions to people who need them. Gloria Gilbert Stoga, president and founder of Puppies Behind Bars, says, "The puppies have affected the lives not only of their puppy raisers, but of virtually all the inmates and staff at the prison." Both puppies and inmates are transformed by working together as a team.

HOSPITAL HOUNDS

Nursing homes are probably the most common place to find therapy dogs working today. One Australian study concluded that, "Like children, many elderly people benefit from something constant, something fixed, such as the undying loyalty of an ever-present pet, as everything else changes around them." Another study found that when nursing home residents spent as little as 30 minutes each week with a dog, their feelings of loneliness were significantly reduced. **Alzheimer's disease** patients who spend 30 minutes per week with a visiting dog become calmer, more responsive, and better able to think clearly. Nursing home staff members have been found to benefit as much as, or even more than, patients do from having a residential dog to help soften and humanize the institutional setting.

As a result of these well-known benefits of introducing therapy dogs into nursing homes, hospitals are now beginning to get with the program. The Prescription Pet Program, a joint venture of the volunteer association of the Denver Children's Hospital and the local veterinary medical society, arranges for trained therapy dogs and their owners to make short visits to consenting patients in their own rooms. The volunteers spend as little as 10–15 minutes with each patient every two weeks, and also make regular visits to specialized areas of the hospital, such as the **dialysis** unit, special care nursery, or psychiatric unit. Because of infection concerns in the hospital setting, all volunteer dogs wear special smocks to reduce **dander** and allergic reactions, are bathed before each visit, and receive frequent evaluation and testing to make sure they carry no bacteria or parasites that could be spread to patients. The program has been a great success. Patients who participate are calmer, more relaxed, and have lower blood pressure than patients who don't interact with the dogs. As one therapy dog volunteer put it, "Dogs don't see what people see. They don't see a broken arm or a missing leg or a scar, which may make a patient embarrassed. Dogs make no judgments. They don't want anything from you and they don't have to say the right thing. They don't expect anything except perhaps a pat. They just want to give love."

HOSPICES NEED THERAPY DOGS, TOO

Animals have a special role to play in hospice care, providing comfort and consolation to terminally ill patients during their last days. One social worker who has studied the dying experience of terminal cancer patients in a nursing home came to believe that therapy dogs and other animals can help patients work through their feelings of anxiety and despair, because the animals' emotions don't get in the way. Human caregivers, on the other hand, must deal with their own fears of death, and so may unconsciously discourage patients from expressing their sadness and pain. Dogs and other pets are recognized as extremely important to support patients who often feel isolated and rejected because of the nature of their illness. As an added benefit, therapy dogs have been reported to reduce stress and emotional exhaustion among caregivers who deal with such patients on a daily basis. These positive results override any potential risk of infection, especially if guidelines developed for infection control in long-term-care institutions are strictly followed.

PAWS (Pets Are Wonderful Support), a San Francisco group providing support services for HIV patients, has published extensive guidelines for minimizing the health risks of contact between these especially vulnerable patients and pets. The Centers for Disease Control and Prevention (CDC) has determined that there is *no* risk of the AIDS virus being transmitted to humans or other animals by pets. This is a specialized, difficult, but potentially rewarding type of work for therapy dogs and their owners.

DOGS GO TO SCHOOL

In both health care and educational institutions, therapy dogs play a role in assisting speech, occupational, and physical therapists to supplement traditional hospital treatments. The dog seems to be a universal motivator for patients and students. Once a dog is involved in therapy, work suddenly becomes play for many people. This is especially true in work with children.

Dogs are even helping kids learn to read in schools and public libraries that have adopted the innovative R.E.A.D. (Reading Education Assistance Dogs) Program, originally developed in Salt Lake City, Utah, by dog/handler teams from the Delta Society's Pet Partners. The idea behind R.E.A.D. is that children who have problems in school, especially those having trouble learning to read well, often suffer from low self-esteem, have difficulty concentrating, and may be intimidated by the idea of reading in front of people. But these same children are often *eager* to read to a friendly dog that is willing to lie quietly beside them and "listen" as they read a story. This is also helpful to children for whom English is their second language, who may struggle to get by in a mainstream classroom. For children who have participated in these dog-centered literacy sessions, reading scores have improved significantly. Improvements have also been noted in self-confidence and self-esteem, attitudes toward school and learning, and overall school grades. Most trained therapy dogs adapt easily to the classroom setting—especially those that love lots of attention and petting, don't mind a noisy room full of kids, and are happy to rest quietly next to a child for as long as needed. The big advantage for students is that dogs never complain or criticize when the kids make a mistake while reading. Instead, they just listen quietly.

Dogs have also been enlisted as "co-therapists" by school counselors, who find that withdrawn children are often much more willing to talk in the presence of a friendly dog. Often, students will come to see the dog and stay to talk to the counselor while they pet and play with the animal. In this way, the counselor is able to interact with many more students than would otherwise be possible.

STILL MORE JOBS FOR THERAPY DOGS

Dogs have been incorporated into substance abuse treatment programs around the country with some success. For example, a halfway house for alcoholics in Bridgeport,

Dogs and Politics

Associating with dogs doesn't just make people happier, healthier, and more relaxed; studies have shown that having a dog by your side also affects the way *other* people see you. "Dog people" are perceived as friendlier, happier, less tense, and less of a threat to others (unless, of course, the pup in question is a snarling attack dog straining on the end of its leash!). The "popularity effect" may explain one of the well-known therapeutic qualities of dogs—the way they often improve people's social interactions and serve as an "ice breaker" to ease awkward moments and improve conversation between strangers.

It also has a lot to do with the way dogs have been used by American politicians to enhance their public image in the eye of ever-present cameras and reporters. President Harry Truman once said, "If you want a friend in Washington, get a dog." Only five U.S. presidents have *not* been dog owners. (Not surprising in our canine-loving country, where such a fundamental lack might plant seeds of mistrust among a large segment of voters.)

From Franklin Roosevelt's beloved Scottie dog Fala, to Richard Nixon's cocker spaniel Checkers, to John F. Kennedy's fluffy little Pushinka (a gift to the Kennedy family from Soviet dictator Nikita Khrushchev), dogs have long been part of the presidential image, for better or for worse. During the 1992 presidential campaign, George H.W. Bush once claimed that his spaniel Millie knew more about foreign policy than either of his two rivals, Bill Clinton and Al Gore.

Connecticut, gave each new guest a puppy to care for. The dogs were given names such as "Brandy," "Cognac," and "Vodka." They went home with discharged residents at the end of the treatment period, to remind the recovering alcoholics of their sobriety training and to give them a sense of purpose in their new lives. At the Abbey alcoholic treatment center in Winfield, Illinois, a mixed-breed dog named "Tramp," injured by a car after being given alcoholic drinks at a West Side Chicago bar, was adopted and became a much-loved member of the staff.

Dogs have been placed with emotionally disturbed children who have been orphaned or placed in foster homes. In caring for the dog, a child is often encouraged to act positively and to think about something besides his or her own personal problems. Dogs can help disturbed children bridge the gap between fantasy and reality, therapists say. The unconditional love of a dog can and does work wonders for a child at a turning point in his or her young life.

Finally, in the post–9/11 world, crisis intervention dogs like those that comforted people in New York after the terrorist attack will likely be utilized more and more. Wherever people are in trouble, sad, or lonely, there is a dog somewhere, waiting to help.

Chapter 4

Horses for Health

HUMANS AND HORSES HAVE SHARED a unique bond at least since the beginning of recorded history. In fact, horses are among only a select few species to have become **domesticated**—a process that gave the horse protection from danger as well as food for survival. In return, humans gained a means of traveling faster than ever before.

Even today, horses remain the subject of myth and ritual. They have been romanticized as warriors, heroes, status symbols, and religious icons. They represent everything from life energy and superhuman power to magic, freedom, and transformation. The way we associate horses with speed and power is still reflected in the names of many popular automobiles: Bronco, Mustang, and Colt, to name just a few.

Despite the fact that the horse is, by nature, a relatively timid animal, captured and tamed by humans, most of us still view horses as all-powerful and mysterious. It's the dual nature of the horse—at once large, strong, and a bit wild, yet gentle enough to allow a puny human to perch upon its back—that sets it apart from other animals.

HOW MODERN THERAPEUTIC RIDING BEGAN

Organized therapeutic riding began in the early 1950s in Scandinavia, after Lis Hartel of Denmark, severely disabled by polio, won a silver medal in **dressage** at the 1952 Olympic Games. This impressed medical and equine professionals alike, and quickly led to the establishment of centers for therapeutic riding across Europe. The idea soon spread to the United States and Canada, and in 1969, the North American Riding for the Handicapped Association (NARHA) was founded. Today, NARHA oversees a growing number of therapeutic riding programs. It provides training, guidelines, and certification programs for those who work in the field.

BENEFITS

The physical benefits of therapeutic riding may include improvements in balance, posture, and walking ability. This works partly because a person's **gait** is, in some ways, similar to that of a horse. The horse allows a disabled rider to experience movement that is natural, rhythmic, and progressive. This, in turn, stimulates and massages the patient's muscles, improving nerve impulses, muscle tone, and posture. Many disabled riders also find that the natural body warmth of the horse helps relax tight muscles. Furthermore, people who support therapeutic riding claim that improvements have been seen in a wide range of patients' bodily functions, including breathing, circulation, and bladder and intestinal function, as well as overall coordination.

Mental and emotional benefits include improved confidence and self-esteem that naturally occurs when a disabled rider learns to control a powerful, 1,000-pound animal. Because the horse *is* so large, but also tends to be nervous—reacting instantly and sometimes dramatically to loud voices or rough handling—emotionally disturbed individuals are often seen to exercise more focus and self-control around horses than in many other areas of their lives. Over time, many such people learn to improve their behavior and functioning in everyday situations, even away from the horse.

THERAPY HORSES ARE SPECIALISTS

Equine-facilitated therapy has evolved into a number of different forms, each with its own set of goals and guidelines. The oldest and best-known kinds of therapy all involve horseback riding in some form, but each has distinct objectives and requirements, and are suited for very different kinds of patients.

Hippotherapy

Hippotherapy is an entirely passive form of riding in which the patient allows the horse to move him or her. Most often, this will involve several trained helpers on the ground: one to lead the horse, and at least two "side-walkers" who walk alongside the rider and prevent falls by holding the rider's body with a steadying hand. Hippotherapy is often used for children with severe posture and movement problems. It has been especially successful with cerebral palsy patients. Through hippotherapy, children born with severe disabilities can make big improvements in muscle tone and in their ability to control their head and upper body. They may also be able to reduce involuntary muscle spasms that often make it difficult for them to coordinate their movements for walking and other activities. The patient's response to the horse's movement is unconscious; it just "happens," automatically, through following the rhythmic movements of the horse as it walks.

Therapeutic Riding

Therapeutic riding is the proper term for a more active kind of exercise on horseback. This includes exercises specially designed to stretch, strengthen, and relax the rider's muscles. The main goal is improved sitting, standing, and walking balance; greater flexibility; better coordination and reflexes; improvements in breathing and heart function; and overall better body control.

Riding for Rehabilitation

After successful treatment with hippotherapy or therapeutic riding, a disabled student may progress to riding for rehabilitation, in which the rider learns to take active control of the horse. Students learn how to sit *correctly* on the horse, and how to give the proper cues in the right sequence. This requires both conscious decision-making and **sensory-motor** skills, and the horse gives its rider instant feedback by responding to what the rider asks, or by *not* responding if the signals are not properly given. Often, disabled riders become so skilled that their disability all but disappears when they are on a horse.

Other benefits are more psychological and social than physical. The riding program itself provides structure and organization, and encourages the student to improve memory and self-control along with controlling the horse. Riders who have especially benefited from riding for rehabilitation include autistic children, as well as those with speech defects and developmental disabilities.

Vaulting

A specialized and very impressive type of therapeutic riding is **vaulting**, in which the student learns to perform gymnastic exercises on horseback (Figure 4.1). This requires skill, practice, and courage on the part of the student, and a quiet, steady horse with comfortable, rhythmic gaits. Although vaulting may be too physically demanding for many disabled students, some young people with delayed development and/or behavioral problems have been amazingly successful in mastering daring, circus-like routines on the back of a moving horse. The benefits depend upon forming mutual trust between the horse and performer. The horse is often a much better disciplinarian than any human teacher could hope to be.

Equine-facilitated Psychotherapy

An exciting new branch of therapy with horses is equine-facilitated psychotherapy, along with a related field often called **equine-assisted activities**, or **equine experiential learning**. These forms of horse therapy don't often involve much riding. Instead, they focus more on the ways patients learn to relate to the horse as another thinking, feeling creature. Sometimes just taking care of horses—talking to them or grooming and feeding them—has been shown to yield surprising psychological and behavioral benefits.

Equine-facilitated psychotherapy is often provided through a team approach. The team will consist of a NARHA-certified riding instructor, a licensed mental health professional, and one or more horses. The patient completes the "team." Sometimes, this kind of therapy will be offered to groups of patients with similar needs. Other times, it will take the form of individual therapy. Occasionally, the riding instructor and licensed therapist will be the same person—often a professional psychologist, psychia-

Disabled Rider Goes for the Gold

Riding a horse is a difficult skill to master. Riding at the top level of international competition is a dream few riders ever achieve. But winning a silver medal at the 2004 Paralympic Games in Athens was a remarkable personal victory for Lynn Seidemann, a 41-year-old Texas woman who had become a **paraplegic** in a skiing accident 20 years earlier. With her horse Phoenix B, a 15-year-old Polish Warmblood who, said Seidemann, "really takes care of his rider," the internationally renowned rider earned a second-place finish in the Individual Freestyle dressage event with an amazing performance.

Seidemann had always been an athlete, but she began riding to rebuild her back strength and improve the spinal curvature that resulted from her accident. It was a mixture of determination and curiosity that led her to see how a person who would normally be confined to a wheelchair could possibly stay on the back of a horse. She also decided to try therapeutic riding because it was the only sport she could find that forced her to use her back and abdominal muscles in the opposite way from what the wheelchair required. The special bond with her horse, she learned, was the best thing of all.

trist, social worker, guidance counselor, or other mental health professional who has a personal interest in and knowledge of horses. At times, the connection between what seem like two vastly different activities—psychology and horseback riding—will come about in the opposite way: A skilled, experienced horseperson with a special interest in human psychology will study to get the credentials needed to provide therapy, with the goal of starting a practice that uses equine-facilitated psychotherapy.

Equine-facilitated psychotherapy has been used to help children, adolescents, and adults—anyone who has severe mental problems or who is seeking help for troubling life issues. A lot of clinical evidence suggests that this kind of therapy can help people suffering from many conditions, including depression, low self-esteem, learning disorders, anxiety, **attention-deficit/hyperactivity disorder (ADHD)**, substance abuse, eating disorders, brain injury, autism, **Tourette's syndrome**, **post-traumatic stress disorder**, and even diseases such as **schizophrenia**. Equine-facilitated psychotherapy is always provided by a trained professional.

Equine Experiential Learning

Equine experiential learning or equine-assisted activities emphasize the *educational* benefits that clients may gain from interaction with horses, without making specific therapeutic claims. The person who runs this kind of program does not need to be a licensed therapist. He or she *will*, however, need to have training to work with both horses and people. Most often, this means the leader of the program will be a qualified therapeutic riding instructor with additional training in equine-centered learning experiences. In equine experiential learning, the basic idea is that clients

Figure 4.1 This young woman is performing therapeutic exercises on horseback.

will learn about themselves through interaction with their environment—including people, animals, the natural setting, and situations arranged in specific ways by the instructor or team of instructors, to bring about particular results.

Some common goals of most equine-facilitated programs are for patients to gain improved self-esteem and self-awareness, develop trust and social skills, and gain a greater ability to make choices. Other objectives are to help participants set goals, improve problem-solving, and develop a strong sense of personal responsibility, respect, and caring for others.

SUCCESS STORIES

One outstanding example of a successful equine therapy program is found at the Green Chimneys School, a residential treatment center in Brewster, New York. Here, equine activities have been used effectively to help children with emotional, behavioral, and psychological disorders. According to Miyako Kinoshita, the equine program director at the school, "Every activity in the horse barn at Green Chimneys is used for learning. Horses, staff, and all the activities in the barn teach academics, life skills, and relationship skills." Children at Green Chimneys work with horses, both mounted and unmounted, and perform basic barn chores as part of the curriculum. Learning and therapy are joined into a program where horses are the focus of both fun and hard work, of academics and important life lessons.

Another well-known program works in association with the Thoroughbred Retirement Foundation (TRF), an organization whose primary goal is to offer safe retirement havens for racing Thoroughbreds after they can no longer race. The nation's largest Thoroughbred rescue operation,

TRF has given new leases on life to hundreds of horses whose lives would otherwise have been in jeopardy. Since 1982, several TRF facilities in New York, Maryland, Kentucky, and Florida have pioneered programs in which retired racehorses are cared for by prison inmates or at-risk youth confined to residential schools as a result of violent or criminal acts. Inmates young and old are given the opportunity to learn useful vocational and life skills through caring for the horses. In the process, many experience trust and bonding for the first time in their lives, and begin to learn ways of interacting that depend more on empathy and understanding than on power and control. The horses, often rescued from abusive situations, begin to heal just as the inmates do.

Equine therapy can also have a competitive side. The centerpiece of international competition for riders with disabilities is the Paralympic Summer Games, featuring dressage, and the Special Olympics, which includes equestrian events at both the national and international levels. Carriage drivers with disabilities also compete with able-bodied drivers in many competitions at every level. For disabled riders who don't wish to compete, special worldwide vacation packages are available, bringing together horses and disabled riders of all languages and nationalities.

GETTING STARTED AS AN EQUINE THERAPIST

To prepare for work in the growing field of therapeutic riding, hands-on experience with horses is a must. Riding lessons, barn or ranch jobs, and participating in horse-related sports are good ways to learn about horses. College courses in health and human service–related fields like psychology, social work, and occupational or recreational therapy, combined with a solid background of knowledge and experience with horses, is the best preparation to become an equine therapist. Special educational seminars and certification courses offered through NARHA or a newer organization called EAGALA (Equine Assisted Growth and Learning Association) are available in specific areas of equine therapy. Formal certification may be required for anyone seeking full-time employment in the field.

THE FUTURE OF THERAPEUTIC HORSEMANSHIP

Hundreds of therapeutic riding programs exist in all 50 states, and there is no lack of inspiring stories and observations suggesting that equine therapy heals. What *is* lacking, however, is sufficient scientific data to support the enthusiastic observations of doctors, therapists, instructors, parents, and riders. The few studies that have been done generally support the idea of therapeutic riding programs. However, much more research is needed to provide credibility to those currently practicing equine therapy; to develop better, more consistent professional training and standards; and to further the establishment and growth of the field as a respected branch of the health professions.

Chapter 5

Fur, Feathers, and Fins

CATS ON CALL

THOUGH THE DOG REMAINS "man's best friend," there's something about cats that makes them the number-one choice for some therapy situations. Cats tend to be quieter than most dogs, for one thing. They're smaller, less pushy, and more often content to simply sit on a person's lap or hospital bed and be gently petted and stroked. And there is one other thing: Dogs just don't purr.

"A cat's purr stimulates our auditory sense and provides us with a peaceful respite from the mechanical noises that are constantly bombarding our senses," according to Dr. Allen Schoen. "In our fast-paced lives," he says, "cats offer us an animal friend, a companion that offers great psycho-social benefits of love and companionship without too many demands."

Maybe that's one reason cats are now the most popular pet in the United States, with nearly 60% of American households including one or more fluffy felines. It's been proven that stroking cats and feeling that vibrating purr can lower a person's blood pressure, help heal heart disease, and reduce stress. Linda Hines, president of the Delta Society, says, "We've seen a very definite increase in the number of cats registered in our national pet-partnership program." Besides the fact that cats just seem naturally soothing to many people, those who have worked with them have found that nursing home patients, in particular, often enjoy grooming and playing with therapy cats. Combing a fluffy coat or buckling a cat's slender collar can be a great way to practice fine motor skills. But mostly, just cuddling a cat helps many people feel less lonely and depressed. It helps them forget about themselves and their illness, at least for a little while. Even agitated patients often become calm and content when they are with a therapy cat.

For many of the same reasons, cats are turning up in therapy programs for children and teens in mental health facilities and group homes. Some facilities keep resident cats and allow patients to help care for them. This promotes responsibility and encourages better concentration, focus, and the ability to follow directions.

Both longhaired and shorthaired cats can be therapy animals. What's most important is that the cat must be calm and should get along well with strangers. A good therapy cat is happy to sit with a person for a long time (Figure 5.1). Often, long-lost memories of Alzheimer's patients will be stimulated during interactions with a therapy cat. Stroking

Figure 5.1 Therapy cats bring smiles to patients' faces.

a cat's soft fur will often trigger a cascade of memories in elderly people who have begun to forget so much of their past—but memories can't be rushed.

Therapy cats must be at least one year old, since older animals are both calmer and less likely to catch diseases from being out with people and other animals. They must be up-to-date on their vaccinations, be certified by a veterinarian, and complete a training program that includes exposure to loud noises, crowds, dogs, and frequent handling. They'll also need to pass a temperament test designed just for cats. Often, but not always, retired show cats—which are already used to crowds, noises, and lots of handling—will make the best therapy cats.

PIGS, PONIES, PARAKEETS—AND MORE

Just as some people keep rare animals as pets, some unusual pets can make great therapy animals. Although the numbers of "alternative" animals in therapy programs are relatively small, their presence looms large to the people whose lives have been brightened by them.

Rabbits and Small Animals

Rabbits are docile and appealing animals, which makes them well suited for therapy work. Oreo, Jessie, and Henry, for example, are three "staff rabbits" with the American Red Cross of Susquehanna Valley in south-central Pennsylvania. Harvey and Chrystal are also volunteers with the group. Owners Cindy Drob and her 11-year-old daughter, Aimee, became Red Cross volunteers, said Drob, "because we have a lot of pets at our house. I know they're therapy for me, to be home with all of these animals . . . and I just wanted to share it with other people."

Small creatures like hamsters, gerbils, ferrets, and even lizards are working as therapy pets, too. The Companionable Zoo, at the Devereux School in the Philadelphia area, keeps all these animals—plus occasional goats, sheep, and miniature horses—in classrooms or in nearby buildings on campus. The children at the school, who all struggle with emotional, behavioral, and educational difficulties, help care for the animals and gardens connected to the zoos. Psychiatrist Aaron Katcher, a pioneer in animal-assisted psychotherapy who works with the Devereux School students and other similar programs, declared the project a success on many levels: "Positive effects include decreases in under-controlled and aggressive behavior . . . improved cooperation with instructors . . . and appropriate behavior."

Even pot-bellied pigs have been trained as therapy animals—their comical appearance brings laughter and joy to many patients. In England, a ferret named Wombat once made an instant connection with a young autistic girl who had never responded to people, dogs, or other animals before. When Wombat was taken out of his box, the girl began to stroke him. She spoke to him and set him gently on the floor; then she got down on the floor herself to crawl along beside the little animal, talking to him all the while. At the end of the visit, the little girl kissed the ferret good-bye and waved to him—things she had never done before.

EXOTIC ANIMALS

Exotic animals can also be therapists (Figure 5.2). At Green Chimneys School in Brewster, New York, one resident farm animal is Angel, the llama. Susan Brooks, a clinical psychologist at the school, recalled one boy who had lost the only caring person in his life—his drug-addicted father.

The boy seemed to be made of stone, refusing to show any emotion over his terrible loss, until the day he sat down next to Angel the llama, buried his head in her thick fur, and sobbed as if he would never stop. Angel stayed perfectly still, allowing the boy to finally express his grief.

Perhaps because they are so pleasant to touch, llamas seem able to connect especially well with autistic children. Darlene Meyer, an occupational therapist who works for a public school district in Oregon, sometimes brings her llamas Steinway and MacCloud to work with special-needs children at the schools. One time, she remembers, an autistic child who had trouble making eye contact began petting MacCloud's side. "He gradually worked his way up toward the front of the llama until he finally looked into those expressive llama eyes," said Meyer. "Everyone in the classroom held their breath so the magic moment wouldn't be broken. Who knows? Perhaps that was a first step toward making eye contact with humans."

Ponies and Donkeys

Two small members of the horse family—donkeys and ponies—are not often ridden but sometimes perform near-miracles with troubled children. Donkeys, compact and cute, share the universal equine trait of wary alertness. Children find that they must *earn* the trust of the donkeys before being permitted to approach and pet them. One boy, Gil, learned from this experience how to soften his approach to people as well, so as not to scare them off with his formerly aggressive attitude.

Personal Ponies is a national organization that provides small Shetland ponies to families of disabled children. Families unable to keep a pony at home are often given one

Figure 5.2 A trained monkey assists a disabled person. Even exotic creatures like this can make effective therapy animals.

to visit and care for, while the pony is boarded with a nearby trained volunteer. When a child outgrows a pony (or, in some sad cases, succumbs to his or her illness), the animal is returned to the program and placed with a new family; since the typical life span of the ponies is about 30 years, each pony may serve many children.

Feathered Friends

Birds may seem to be unlikely therapy animals, but caring for caged birds has brightened the lives of many prisoners over the years. Robert Stroud, the legendary Bird Man of Alcatraz, was a violent convict kept in isolation from other inmates. He became a productive citizen by learning about and caring for canaries that visited the prison exercise yard. The story of a caged man becoming fascinated with caged (and uncaged) birds later inspired numerous books and an Academy Award–nominated movie. Since the mid-1970s, a number of prisons and institutions for the criminally insane have introduced mascot programs in which inmates are given the opportunity to care for small animals and fish.

The *Real* Birdman of Alcatraz

Robert Stroud who was portrayed by Burt Lancaster in the 1962 Academy Award–nominated film *The Birdman of Alcatraz* as a mild-mannered, nature-loving tragic figure, was, in fact, a violent, unpredictable criminal who spent nearly all of his adult life in prison for committing more than one brutal murder. Although probably the most famous inmate ever to live at Alcatraz, Stroud actually began and ended his involvement with birds while at Leavenworth Federal Penitentiary in Kansas, where he was imprisoned for 30 years before serving time at Alcatraz. Held in solitary confinement during most of his time at Leavenworth due to his frequent threats and attacks on other inmates, he developed a keen interest in birds after finding an injured sparrow in the prison recreation yard.

Stroud was permitted to breed and raise nearly 300 birds and maintain a research lab inside two adjoining segregation cells, as a way to make productive use of his ample time. As a result of his direct observations and intensive study, he became a leading authority on canaries and their diseases. Though he had come to prison with only a third-grade education, he authored two books on birds, and developed and marketed medicines for various bird ailments. But after several years, prison officials shut down his operation when they discovered that he was using scientific equipment to make whiskey.

He died in a prison hospital in 1963, without ever seeing the movie that portrayed him in such a flattering light.

Birds have proven especially popular and useful. One reason may be that some birds have been shown to bond with their keepers, to express emotions like jealousy and joy, and to be highly intelligent companions.

Fishy Friends

Last, but not least, are fish. Although an unfriendly person is sometimes referred to as a "cold fish," this expression is something of a misnomer. Although it is true that fish are **cold-blooded**, many studies have found that aquarium fish can have a beneficial effect on people who watch them swim gracefully in their tank. That is why doctors' and dentists' offices sometimes feature an aquarium in the waiting room, to help nervous patients relax. A study conducted in the 1980s concluded that watching fish decreased a person's blood pressure to below the level of someone sitting comfortably in a chair and simply resting. Watching fish was also found to produce a state of calm relaxation, and to be especially beneficial for elderly people.

Fish are the mostly unsung heroes of the animal therapy world. They prove that bigger isn't always better, and that even though the dog may still be man's best friend, that circle of friends is almost as wide as the entire animal kingdom.

Chapter 6

Doctor Dolphin

WHEN DEENA HOAGLAND FIRST BROUGHT her 3-year-old son, Joe, to swim with dolphins, she never guessed that the outing she had planned on a whim would change Joe's life, and the lives of his entire family, forever. Joe had suffered a **stroke** during his third open-heart surgery to correct a birth defect. The stroke had left him partially paralyzed, unable to use the left side of his body much at all. So far, he hadn't responded well to traditional therapies. Doctors had urged the Hoaglands to place Joe in a full-time rehabilitation center, but his mother, as a licensed clinical social worker and teacher, believed she could help him more on her own.

The family had just moved from Denver, Colorado, to Key Largo, Florida, when Hoagland learned about Dolphins Plus, a nearby research center specializing in dolphin communication. Knowing that water was considered one of the best therapeutic environments for helping people regain lost mobility, Hoagland arranged to bring Joe to the center, little knowing what to expect. By then, she was getting desperate. Since his stroke, Joe rarely smiled or laughed anymore; he seemed listless, withdrawn, and depressed. As a mother, Hoagland longed to see her son play and laugh again, if nothing else.

She could never have guessed that Joe would befriend a dolphin named Fonzie that day, much less that this would be the long-sought breakthrough on the road to recovery for her son. The toddler smiled and giggled for the first time since his stroke when he splashed in the water with Fonzie. As he continued to play with the dolphins, Joe soon began to walk and behave like a normal little boy. Along with dramatic improvements in his muscle tone and flexibility, he regained his self-esteem and energy. After 22 months of play sessions with Fonzie, while also participating in swim therapy and traditional physical therapy, Joe had come close to a full recovery and was soon able to go to regular school instead of the special-needs preschool he had attended before Fonzie changed his life.

The next year, Deena Hoagland founded Island Dolphin Care to help children with special needs—developmental and/or physical disabilities, emotional problems, and critical, chronic, or terminal illness—and their families. She remembers thinking, "If dolphins could help Joe feel better about himself, and motivate him to try new tasks, then the dolphins might also help others." Now Hoagland and three other full-time therapists work with thousands of children and their families each year.

WHAT IS IT ABOUT DOLPHINS?

Why dolphins? Just as with horses, the answer lies in the unique nature of the animal. Dolphins have held a special fascination for humans throughout the ages—in fact, dolphins are prominently featured in several ancient cultures and religions.

Another reason dolphins are so deeply rooted in the human imagination is their undeniable intelligence. Researchers have determined that even humans may rank second to dolphins in brainpower—at least in terms of brain size and anatomy. Still another attraction is dolphins' highly refined communication system—so different from our own but perhaps equally expressive and complex. A dolphin produces clicking sounds from just below its blowhole. When the sounds hit an object, they bounce back and tell the dolphin where the object is. Dolphins also let out a range of whistles and squeals that distinguish a particular dolphin from other dolphins, and convey emotional states and other information.

With long (8 to 10 feet [2.4 to 3 meters] for an adult), streamlined bodies, an average weight of 400 pounds (181 kg), swimming speeds of up to 35 miles (56 km) per hour, and as many as 26 razor-sharp teeth in each side of their jaws, dolphins are gentle, graceful, and friendly-looking—all traits that humans find comforting and appealing. The delicate, curved shape of dolphins' jaws gives them a constant toothy grin. What's more, most dolphins seem to be intensely interested in humans.

DOLPHIN-ASSISTED THERAPY

Just being in water offers many therapeutic benefits for people with various types of disability. Its **buoyancy** can temporarily free a person who must spend most of every day in a wheelchair, and a large volume of water provides a constant surrounding pressure that can be soothing to someone with chronic pain and hypersensitive nerve endings. But advocates of dolphin-assisted therapy claim a whole range of additional benefits for this special form of treatment.

Some people who are thought to benefit from swimming with dolphins are those with physical disabilities such as spinal cord injuries, cerebral palsy, complications from strokes or brain damage from accidents, chronic diseases like **diabetes** or **multiple sclerosis**, or terminal illnesses, including various types of cancer. Others suffer from psy-

They Call Him Flipper

The most famous dolphin of our time was known by a stage name. Her *real* name—Mitzi—would probably have surprised the many fans of Flipper, but the fact that "he" was a she might have shocked them even more. Mitzi was the first pupil of Milton Santini, an early 1960s pioneer in dolphin training and the owner and operator of Santini's Porpoise School in Grassy Key, Florida. As Santini's star student, Mitzi was chosen to star as Flipper in the original 1963 movie of the same name that was filmed at the school. Although Mitzi provided the engaging dolphin smile that won hearts and inspired two long-running television series (1964–1968 and 1995–2000), it was a male stunt dolphin named Mr. Gipper who performed Flipper's famous tail-walking maneuver—Mitzi could never quite master that trick.

chological and emotional disabilities: autism, depression, or other mental illnesses; attention-deficit/hyperactivity disorder; or **Down's syndrome**, for example. Some are victims of physical and sexual abuse; many are children, who seem to have a natural affinity for animals. But adults, too, are said to have benefited from dolphin-assisted therapy. Many people report that they feel happy and hopeful after swimming with dolphins, and accounts of dramatic, long-lasting improvements in emotional well-being aren't hard to find.

HOW DOES IT WORK?

Health benefits from dolphin-assisted therapy are supported by measurements of brain wave patterns, blood chemistry, brain scans, and cell analysis of patients. But ways to explain these improvements are still being studied. Explanations range from the idea that swimming with dolphins somehow stimulates the immune system to promote healing, to the more romantic notion that patients who interact with dolphins experience such joy that they become more "open" to recovery.

Some people believe that dolphins are uniquely sensitive to the needs of people with disabilities, and seek to help them through playful expressions of concern. There is a "secret language," they say, shared by dolphins and people with disabilities. Perhaps dolphins communicate both with sound and with a variety of movements, and are extra-sensitive to the body cues of people who have difficulty communicating with others through speech, such as those with autism or developmental delays (Figure 6.1). Some say it's almost as if the dolphins can understand the thoughts and actions of people who are unable to put these things into words.

ETHICAL AND PRACTICAL ISSUES

As you might expect, dolphin-assisted therapy is quite expensive. The care, feeding, and training of creatures as large and complex as the dolphin require special facilities and skilled caretakers. Besides that, some people believe that using dolphins in this way exploits them, and there are some indications that captive dolphins do suffer from a range of stress-induced physical and mental symptoms.

That's why some scientists are hard at work creating a "virtual dolphin experience," which they hope will capture the joy and freedom of swimming with the animals using video and computer technologies, eliminating any need for real, live dolphins. David Cole, a computer scientist living in the Los Angeles area, heads a group of computer wizards, doctors, and naturalists called the AquaThought Foundation. He is one of several people currently working

Figure 6.1 As this child can attest, there's nothing like swimming with dolphins.

to develop computer-simulated dolphin experiences. Cole's prototype, called Cyberfin, is already up and running. It is a "virtual reality interaction" that simulates swimming with dolphins. Once it's perfected, Cole believes the system will be a boon both to humans who can't afford a live dolphin swim and to the wild dolphins that will be spared a life of captivity for the purposes of human therapy and recreation.

Chapter 7

Is Animal Therapy in Your Future?

INTEREST IN THE HEALING POTENTIAL of animals is growing worldwide, both in the scientific community and in popular culture. So what does the future hold for the fascinating and unique field of animal-assisted therapy?

Research and clinical observations over the past three decades have shown a scientific basis for the centuries-old belief that the association of people with animals and the natural environment contributes to the overall health and well-being of humans. The jury is still out on whether this association is as beneficial for the animals as it is for the humans with whom they bond. Some researchers warn that using essentially wild animals like dolphins or monkeys for human healing may be more stressful to the animals than was once thought.

But most of the domestic animals commonly accepted into animal-assisted therapy programs seem to have a genuine affection for humans and a love for their work. With many studies showing specific, measurable medical benefits from animal therapy, such as increased survival rates for heart patients; reduction in blood pressure and stress levels even in healthy subjects; lowered cholesterol levels; improved strength, coordination, and balance in physically disabled riders; and fewer doctor visits among the elderly who enjoy regular contact with animals, AAT is now more than just a theory. At long last, animal healers are gaining real recognition within the health-care professions.

THE FUTURE OF ANIMAL-ASSISTED THERAPY

As the ancient practice of animal-assisted therapy moves into the 21st century, new ways to use the unique healing powers of animals are being explored. While dolphin therapy researchers continue to refine computer-generated simulations of the dolphin experience, a special **neuromapping** system for the collection of brain wave data of subjects before and after interaction with dolphins has already been developed to identify the neurological and psychological factors that may contribute to the benefits of dolphin-centered therapy.

Owners of therapy dogs will also find new ways to use their animals' unique talents to help people in need. Although animals have long provided comfort and assistance in the face of disasters, the tragic events of September 11 dramatically showed the need for highly trained dogs not only to search for victims, but also to provide support and comfort to rescuers and family members (Figure 7.1). More research is needed to deepen our understanding of the role therapy dogs can play in disaster situations—not only to develop a model for expanding this vital service, but also to set up guidelines to protect the health and well-being of the brave and intelligent dogs trained for this kind of work.

Other exciting areas of research in AAT include the possibility of using trained dogs that have the remarkable ability to detect the onset of epileptic seizures and alert their human companions in time for them to take precautions to protect their own health and safety. More research is needed to find out if this is a skill most dogs could learn, or whether only certain dogs are naturally gifted with the ability. Similar skills have been observed in dogs that can predict the onset of a **hypoglycemic**, or low blood sugar,

People and Animals: Exploring the Ties That Bind

A number of research centers have recently been established to learn more about the human-animal bond. One such place is the University of Pennsylvania's Center for the Interaction of Animals and Society (CIAS).

Some projects now under way at CIAS include the Pet Vet Visitation Program, designed to lift the spirits of ill children and their families staying at the nearby Ronald McDonald House; an interactive workshop exploring the relationship between animal abuse, child abuse, and domestic violence; a therapy dog certification program to help students and staff at the vet school get their dogs certified for work in animal-assisted therapy; and an educational program for high school students on careers in biomedical research and laboratory animal medicine.

reaction in their human companions. Even more amazing, some dogs have proven adept at sniffing out cancerous tumors present in the bodies of people, long before doctors or patients had any idea that a tumor existed. Ongoing research will determine how reliable the findings of these tumor-sniffing dogs prove to be compared with more traditional diagnostic tools, and again, whether or not most dogs are born with this potential ability that could be developed with training.

Animal-assisted therapy is moving from a one-time fringe movement into the mainstream, but a lot more rigorous, scientific study of the actual outcomes of various patient-animal interactions is still needed. At the same time, standardized, accredited training programs must be developed to provide the knowledge and skills required to enter this growing discipline. Then animal-assisted therapy will finally take its rightful place as a legitimate branch of the healing arts.

BECOMING AN ANIMAL THERAPIST

If you want to become an animal therapist, the first thing to consider is whether you will want to do animal-assisted therapy as a full-time career, a part-time vocation as one aspect of a larger career, or perhaps simply as an unpaid volunteer. Because AAT is a relatively new field, there is not yet an established "career track" that everyone can or must follow. If you have an independent spirit and enjoy finding your own path in the world, though, there are many ways to learn about AAT, and to prepare for a career that employs the basic standards and practices that have been developed by pioneers in the field for the past several decades.

Any person who wants to enter the field of animal-

Figure 7.1 Trained dogs, like these at Ground Zero after the September 11, 2001, attacks on the World Trade Center, are a vital part of many disaster rescue teams.

assisted activities or animal-assisted therapy must be versatile and interested in pursuing a well-rounded education. The more knowledge you can gain about human and animal behavior and health, the better prepared you'll be to contribute to the field of AAT and to help the people who need it most. Although that can mean a formal college or graduate school degree, there are many opportunities for high school students and graduates willing to seek out other kinds of specialized training.

Start by looking at your own strengths and weaknesses (your school guidance counselor may be able to help you with this). Chances are, if you're interested in AAT, you already know that you love animals, and the idea of using your own positive experiences with them to help other people sounds great. But what kind of people will you help? Before venturing into the field, think about the different people who can be helped with animal-assisted therapy, and about how you might best contribute. Be honest with yourself regarding your comfort level in being around people who may be very different from you or from anyone you've known before.

Do you feel comfortable being around elderly people, or are you more at home with young children? How about people who are mentally or physically disabled, or emotionally disturbed? Would you feel comfortable working with prisoners, or with terminally ill people in hospice care? If you're thinking of volunteering with an animal you already own, keep in mind that your dog or cat may be happier in some places than in others, just like you. It's a good idea to try out a variety of animal therapy experiences to see what level of commitment and what types of programs work best for you and your animal.

HOW TO GET STARTED

There are several different ways to enter the field of AAT. You can get yourself and your pet certified by one of several national credentialing organizations, including the Delta Society, Therapy Dogs International, Therapy Dogs Incorporated, or through the American Kennel Club's Canine Good Citizen Test, and then begin volunteering on your own locally at a nursing home or other institution. In general, certification involves passing both a knowledge test and an animal-handler performance test. Once you're certified by a national organization, you can feel confident in approaching a nursing home administrator and offering your services, and the institution can feel comfortable that you and your pet are qualified for the job you're offering to do. Standards and screening processes for the different certifying organizations vary somewhat and are updated from time to time, so you'll need to follow the guidelines of the particular organization you choose for certification. The institution you approach as a volunteer may have its own requirements, too, such as more frequent health screenings for animals or specialized training for human volunteers.

Once you and your pet are certified, you might want to join and volunteer with an existing animal therapy group in your local area. Many AAT visitation programs are run by local humane societies, SPCAs, and other animal welfare groups. If you live in or near a big city, you may be able to find full-time organized agencies providing AAT services to the community. If there's a college of veterinary medicine near you, check to see whether it includes a human-animal bond department or center. Often, such programs offer community outreach services such as humane education, pet loss grief counseling, and hospice care, among other things. You may be able to get started by volunteering in one of these programs.

You can try contacting local animal shelters, breed rescue groups, or activities directors at local nursing homes to offer your services. By now, many nursing home residents are accustomed to animal visitors, so it may be easiest to get started there. With experience, though, you might try branching out into hospitals, hospices, rehabilitation centers, residential treatment centers, special education schools, prisons, or just about anywhere there are people who could benefit from contact with a friendly animal.

Volunteering is a great way to find out if you want to make a career-level commitment to AAT. If you decide the answer is yes, you'll want to consider furthering your education to prepare for entry into one of the traditional health or human service–related professions. That way, you'll eventually have the credentials needed to work as a professional animal therapist, or to incorporate animals into your career as a counselor, therapist, social worker, activities director, or teacher. Knowledge of animal training, of course, would also be extremely useful for anyone interested in starting an interdisciplinary career that incorporates human health and animal-related activities. You may even dream of starting your own AAT program some day, working within your local community to bring animal-assisted activities and therapy to those who need it most. With knowledge, experience, and a commitment to your work, this is a dream that could very well come true.

Alzheimer's disease: A degenerative disorder that affects the brain and causes a loss of mental capacities, especially later in life.

animal-assisted activity: Activity that provides opportunities for motivational, educational, and/or recreational benefits to enhance quality of life. May be delivered by a trained professional or a volunteer.

animal-assisted therapy (AAT): Goal-directed therapy in which an animal is an integral part of the treatment process; delivered and/or directed by a trained professional.

animal therapist: A person who works with specially trained animals to provide help to people with special needs.

anthropology: The scientific study of the origin, behavior, and development of human beings.

asylum: Out-of-date and offensive term for an institution that houses people with mental disorders.

attention-deficit/hyperactivity disorder (ADHD): A syndrome usually affecting children, characterized by impulsiveness, hyperactivity, and short attention span, which often leads to learning disabilities and various behavioral problems.

autism: A disturbance in psychological development in which use of language, reaction to stimuli, interpretation of the world, and the formation of relationships are not fully established.

bioterrorism: The use of disease-carrying organisms in a terrorist attack.

buoyancy: The tendency of a liquid or gas to cause less dense objects to float or rise to the surface.

cerebral palsy: A condition caused by brain damage around the time of birth and marked by lack of muscle control, especially in the limbs.

chronic: Lasting for a long period.

cold-blooded: Used to describe an animal with an internal body temperature that varies according to the temperature of the surroundings.

coma: A prolonged state of deep unconsciousness.

dander: Tiny particles or scales that are shed from the feathers, hair, or skin of various animals.

diabetes: A medical disorder in which the body cannot properly process sugars.

dialysis: The process of filtering the accumulated waste products of metabolism from the blood of a patient whose kidneys are not functioning properly.

domesticated: Raised or bred to meet human needs.

dorsal: Related to or situated on the back of the body.

Down's syndrome: An inherited disorder, caused by the presence of an extra 21st chromosome, in which the affected person has mild to moderate mental retardation, short stature, and a flattened face.

dressage: A competitive event in which horse and rider are judged on the elegance, precision, and discipline of the horse's movements.

epileptic: Having a medical disorder called epilepsy, involving episodes of abnormal electrical discharge in the brain and characterized by periodic sudden loss or impairment of consciousness, often accompanied by seizures.

equine experiential learning: An activity in which participants gain self-knowledge through interaction with horses.

equine-facilitated therapy: Type of animal-assisted therapy delivered by a team made up of a specially trained horse and a certified therapeutic riding instructor working with or as a licensed mental health professional.

euthanize: To kill an incurably ill or injured animal or person to relieve suffering.

evolutionary biology: The scientific basis by which the development of life on Earth is understood and studied.

gait: A way of walking, running, or moving.

hippotherapy: Type of equine-assisted therapy in which the patient passively allows the horse to move him or her.

hospice: A residence for terminally ill patients where treatment focuses on the patient's well-being rather than a cure.

hypoglycemic: Having an abnormally low level of glucose (sugar) in the blood.

hypothesis: A statement that is assumed to be true, used as a basis for further investigation.

interdisciplinary: Involving two or more academic subjects or fields of study.

mammal: A class of warm-blooded vertebrate animals that have, in the female, milk-secreting organs for feeding the young.

multiple sclerosis: A serious progressive disease of the central nervous system, thought to be caused by a malfunction of the immune system.

neurological: Involving the nervous system.

neuromapping: A process of determining which parts of the brain are responsible for specific activities.

paraplegic: Someone who is paralyzed (unable to move) below the waist.

physiological: Relating to the way living things function, rather than to their shape or structure.

post-traumatic stress disorder: A condition that may affect people who have suffered severe emotional trauma, and may cause sleep disturbances, flashbacks, anxiety, tiredness, and depression.

practitioner: A person who works in a particular profession.

psychological: Relating to the mind or mental processes.

psychology: The scientific study of the human mind and mental states, and of human and animal behavior.

psychotherapist: A professional who treats mental or emotional disorders through the use of psychological techniques designed to encourage communication of conflicts and insight into problems.

reincarnation: In some systems of belief, the cyclical return of a soul to live another life in a new body.

schizophrenia: Psychotic disorder characterized by withdrawal from reality, illogical patterns of thinking, delusions and hallucinations, and accompanied in varying degrees by other emotional, behavioral, or intellectual disturbances.

sensory-motor: Relating to the combination of the functions of sensation and movement.

service dog: A dog specially trained to perform a particular service for a disabled person.

stroke: A sudden blockage or rupture of a blood vessel in the brain, resulting in symptoms such as loss of consciousness, partial loss of movement, or loss of speech.

subjective: Based on a person's opinions or feelings rather than on facts or evidence.

therapeutic: Able to restore or maintain health.

therapy animal: An animal used to help treat disease or restore a person's health.

Tourette's syndrome: A severe neurological disorder characterized by involuntary facial and body movements, grunts, and other noises.

vaulting: Gymnastic exercises on horseback.

Books

Arkow, Phil. *Animal-Assisted Therapy and Activities: A Study, Resource Guide and Bibliography for the Use of Companion Animals in Selected Therapies.* Stratford, NJ: Phil Arkow, 2004.

Beck, Alan, and Aaron Katcher. *Between Pets and People: The Importance of Animal Companionship.* West Lafayette, IN: Purdue University Press, 1996.

Budiansky, Stephen. *The Covenant of the Wild: Why Animals Chose Domestication.* New York: William Morrow & Co., 1992.

————. *The Nature of Horses.* New York: The Free Press, 1997.

Crawford, Jacqueline J., and Karen A. Pomerinke. *Therapy Pets: The Animal-Human Healing Partnership.* Amherst, NY: Prometheus Books, 2003.

Fine, Aubrey, ed. *Handbook on Animal-Assisted Therapy: Theoretical Foundations and Guidelines for Practice.* San Diego: Academic Press, 2000.

Graham, Bernie. *Creature Comfort: Animals That Heal.* Amherst, NY: Prometheus Books, 1999.

Serpell, James. *In the Company of Animals*, 2nd ed. New York: Basil Blackwell, 1996.

Thomas, William H. *The Eden Alternative: Nature, Hope and Nursing Homes.* Columbia, MO: University of Missouri, 1994.

Wilson, Cindy C., and Dennis C. Turner, eds. *Companion Animals in Human Health.* Thousand Oaks, CA: Sage Publications, 1998.

Yates, Elizabeth. *Skeezer: Dog With a Mission.* New York: Harvey House, 1973.

Articles

Bock, Betty Jo. "Second Chances: Horses vs. Criminal Behavior." *EFMHA News* 7 (2003).

Kinoshita, Miyako. "Teaching the Children All Day, Everyday." *EFMHA News* 6 (2002).

McDonnell, Sue. "History of the Horse." *The Horse.* March 2005.

Websites

Abrahms, Sally. *Holistic online.* Available online at *http://www.holisticonline.com/Pets/pet-therapy-bjs.htm.*

Bernstein, Jessica. "Animal Assisted Therapy Brings Love and Companionship." *American Red Cross.* Available online at *http://www.redcross.org/news/co/features/020314animal.html.*

Blow, Richard. "Dr. Dolphin." *MotherJones.com.* Available online at *http://motherjones.com/news/feature/1995/01/blow.html.*

Curtis, Patricia. "The Healing Power of Animals." *Spirituality & Health: The Soul/Body Connection.* Available online at *http://www.spiritualityhealth.com/newsh/items/article/item4449.html.*

Delta Society. Available online at *http://www.deltasociety.org.*

Goodman, Cindy Krischer. "Swimming with Dolphins." *The Miami Herald.* Available online at *http://aegis.com/news/mh/2004/MH041106.html.*

Matthews, Cara. "Green Chimneys: For Anger Control, Visit the Farm." *The Journal News.com.* Available online at *http://www.nyjournalnews.com/rtc/rtc062302_06.html.*

Moore, Arden. "How Cats, the Most Popular Pet in the United States, Climbed to the Top." *Catsecrets.com.* Available online at *http://catsecrets.com/greatpets.htm.*

Moser, Don. "All the Presidents' Pooches." *Smithsonianmag.com.* Available online at *http://www.smithsonianmag.com/smithsonian/issues97/jun97/kennedy_jpg.html.*

National Association of Riding for the Handicapped. Available online at *http://www.narha.org.*

Puppies Behind Bars. "A New Leash on Life." Available online at *http://www.puppiesbehindbars.com/overview.htm.*

Roadside America. "Flipper!" Available online at *http://www.roadsideamerica.com/pet/flipper.html.*

Turner Classic Movies. "Birdman of Alcatraz (1962)—The Essentials." Available online at *http://alt.tcm.turner.com/essentials/essential/fea_birdman.html.*

The Westminster Kennel Club. Available online at *http://www.westminsterkennelclub.org.*

Organizations

Center for the Interaction of Animals and Society
Director: James A. Serpell, Ph.D.
School of Veterinary Medicine, University of Pennsylvania
3900 Delancey St.
Philadelphia, PA 19104

The Delta Society
580 Naches Ave. S.W. #101
Renton, WA 98055
www.deltasociety.org

North American Riding for the Handicapped Association (NARHA)
P.O. Box 33150
Denver, CO 80233
narha@narha.org

Therapy Dogs Inc.
P.O. Box 5868
Cheyenne, WY 82003
therapydogsinc@qwest.net

Therapy Dogs International, Inc.
88 Bartley Square
Flanders, NJ 07836
www.tdi-dog.org

Books

Burch, Mary R. *Volunteering With Your Pet: How to Get Involved in Animal-Assisted Therapy With Any Kind of Pet.* New York: Howell Books, 1996.

———. *Wanted: Animal Volunteers.* New York: Howell Book House, 2002.

Coudert, Jo. *The Good Shepherd: A Special Dog's Gift of Healing.* Salt Lake City: Andrews McMeel Publishing, 1998.

Crawford, Jacqueline J., and Karen A. Pomerinke. *Therapy Pets: The Animal-Human Healing Partnership.* Amherst, NY: Prometheus Books, 2003.

Duncan, Susan. *Joey Moses.* Seattle: Storytellers Ink, 1998.

Garfield, James. *Follow My Leader.* New York: Puffin, 1994.

Hubbard, Coleen. *One Golden Year: A Story of a Golden Retriever.* New York: Apple, 1999.

Nock, Mickey, Anne B. Nock, and H. Marie Suthers-McCabe. *Lapdog Therapy: My Journey From Companion Dog to Therapy Dog.* Onancock, VA: Pickmick Publishing Company, 2002.

Ogden, Paul W. *Chelsea: The Story of a Signal Dog.* Boston: Little, Brown, & Co., 1992.

Singer, Marilyn. *A Dog's Gotta Do What a Dog's Gotta Do: Dogs at Work.* New York: Henry Holt and Co., 2000.

Vinocur, Terry. *Dogs Helping Kids with Feelings.* New York: PowerKids Press, 1999.

Websites

Animal Therapy Net
http://www.animaltherapy.net/

Delta Society
http://deltasociety.org

Island Dolphin Care
http://islanddolphincare.org/nonflash/marinemammals.htm

North American Riding for the Handicapped Association
http://www.narha.org

ABOUT THE AUTHOR

KAY FRYDENBORG is a writer and journalist with a special interest in animal issues. She is the author of *They Dreamed of Horses*, a book about women with unusual horse-related careers, as well as many articles in national horse publications. Her nonfiction story about the history of horses and horseshoeing, "Forging a Life," won the Riverteeth 2002 Prize for Narrative Reportage. She lives on a small farm in Pennsylvania with her husband, three dogs, and two cats. When time and weather permit, she enjoys riding her three horses and competing in horse shows.